Emna Zouaoui
Meriem Bouachari
Chaima Hocine

Characteristic study of an HDPE-based composite material

Emna Zouaoui
Meriem Bouachari
Chaima Hocine

Characteristic study of an HDPE-based composite material

and prickly cactus leaf powder (PFCE)Biodegradable HDPE/Spiny Cactus Leaf Composites

Imprint
Any brand names and product names mentioned in this book are subject to trademark, brand or patent protection and are trademarks or registered trademarks of their respective holders. The use of brand names, product names, common names, trade names, product descriptions etc. even without a particular marking in this work is in no way to be construed to mean that such names may be regarded as unrestricted in respect of trademark and brand protection legislation and could thus be used by anyone.

Cover image: www.ingimage.com

This book is a translation from the original published under ISBN 978-620-6-70545-1.

Publisher:
Sciencia Scripts
is a trademark of
Dodo Books Indian Ocean Ltd. and OmniScriptum S.R.L publishing group

120 High Road, East Finchley, London, N2 9ED, United Kingdom
Str. Armeneasca 28/1, office 1, Chisinau MD-2012, Republic of Moldova, Europe
Printed at: see last page
ISBN: 978-620-3-69572-4

SUMMARY

Composites are constantly evolving towards cheaper, more resistant products, with the added aim of protecting the environment and public health. For this reason, we wanted to discover new bio-composites consisting of a matrix of high-density polyethylene and a filler of prickly cactus leaf powder. When we included prickly cactus leaf powder in HDPE powder at different percentages 5%, 15%, 25%, we studied some characteristics of the mixture.

THANK YOU

*First of all, we thank **God** for giving us the strength and willpower to complete this work.*

*We would like to extend our warmest thanks to **our parents** for all their efforts and support. sacrifices they have made to see us succeed.*
*We would also like to thank our supervisor Mrs **Zouaoui Emna** and our co-supervisor **Gouasmia Abir** for their help.*

*We would also like to thank all the members of the **2022-2023 Master 2, Polymer Engineering** class for their presence and for the happy times we spent together.*
A big thank you to all the members of the jury for agreeing to give up some of their precious time to share their knowledge.

*We would also like to express our gratitude and thanks to all the people who helped us during our training period in the laboratory of the **POLYMED CP2K** unit **in SKIKDA.***
Finally, we would also like to thank all the people who participated in any way in the realisation of this project.

Thank you all...

DEDICATION

Thanks to God Almighty who gave me the courage, the will and the strength to carry out this thesis.
I dedicate this modest work
To my dear parents for their patience, love, support and encouragement
To my sisters-in-law To my brothers-in-law
To my friends and comrades.
Not forgetting all the teachers, whether in primary, middle, secondary or higher education.
Chaima

"With a heart full of love and pride, I dedicate this modest work to you.
To my ideal father Aziz
My precious offering from the god who sacrificed all his means and possibilities to achieve my success. Could you find in this work the fruit of all your pains and all your efforts.
To my beautiful mother Nacira
The inexhaustible source of tenderness, patience and sacrifice. The woman who suffered without letting me suffer, who never said no to my demands and who spared no effort to make me happy.

To my dear brothers Bilel, Mohamed Lamine, ABD EL Raouf, Youcef and Ismail my lovely sister Yasmina you are my dearest people in the world.
Not forgetting my partner Chaima, who has been my companion on this educational adventure. Sharing these moments of study and work was a memorable experience and a sincere friendship.
MERIEM

TABLE OF CONTENTS

INTRODUCTION GENERAL

Over the last few decades, plastics have become a major part of everyday life. Unfortunately, the easy degradation of these materials, which are derived from petroleum, an exhaustible natural resource, has increasingly become a major problem [1-2]. At present, the need for suitable materials is growing all the time, forcing people to combine two or three different components to create new products with improved properties. In fact, the combination of components with complementary characteristics gives rise to attractive properties that are indispensable in certain applications. The materials obtained in this way are called "composites". They are currently a key area of scientific research [3,4]. Plant fibre composites are currently in great demand in a number of sectors, particularly packaging, because of their biodegradability.

Composite materials are, by definition, a combination of two or more components with different structures, enabling a certain number of properties to be obtained that are superior to those of each component on its own. They consist of a reinforcement and a polymer matrix. This combination plays an important role in improving various mechanical, chemical and physical properties, etc. Among the thermoplastic polymers used in composites is high-density polyethylene (HDPE), which belongs to the polyolefin family and has established itself in many applications thanks to its intrinsic properties, which are constantly being improved by the development of new manufacturing processes [5].

Plastics are stable materials that decompose very slowly in nature (around 400 years). The incineration of these petrochemical plastics is also highly polluting, releasing huge quantities of CO_2 and other toxic gases that are harmful to the environment and human health. What's more, the increasing use of these composite materials is creating problems for the management of the resulting waste.In this end-of-study project, we prepared composite materials based on the thermoplastic polymer "high-density polyethylene (HDPE)" reinforced with a filler of prickly cactus leaf powder (PFCE) with different percentages. The main objective of this work is to prepare a new biomaterial with good characteristics and very compatible with the environment in order to minimise the rate of waste as much as possible and achieve a revolution in the field of composites.

This work is divided into five chapters as follows: The first chapter presents the definition of a PE, its chemical structure and its classification. Secondly, this chapter contains a detailed description of the main synthesis processes for high-density polyethylene (HDPE) and its properties.

The second chapter is devoted to a literature review covering general information on composite materials, their classification, advantages and disadvantages.In the third chapter we present an overview of natural fibres, in particular plant fibres, their classification, structure and chemical composition, properties and applications.

In the fourth chapter we mentioned the different materials used in the this study. On the other hand, we have presented the raw material preparation process, as well as the processing protocol used in the manufacture of the composites. This chapter also includes the main methods used to characterise the composites prepared. The final chapter contains the various results obtained and a detailed discussion. Finally, we offer a general conclusion

References

[1] D. C. Miles and J. H. Briston, Technologie des polymères, Dunod, Paris, 1968.

[2] P. Combette and I. Ernoult, Physique des polymères, Tome 1, Hermann, Paris, 2005.

[3] S. Nekkaa, M. Guessoum and N. Haddaoui, Water absorption behavior and impact properties of spartium junceum fiber composites, Int. J. Polym Mater, Vol. 58, 2009, pp. 468-481.

[4] H. Kim, J. Biswas, and S. Choe, Effects of stearic acid coating on zeolite in LDPE, LLDPE, and HDPE composites, Polym. Korea, Vol. 47, 2006, pp. 3981-3992.

[5] S. Meriem. "Elaboration and characterisation of hybrid HDPE/recycled fibre composites. /Organophilic montmorillonite: Study of the effects of the composition and surface treatment of PET fibre". Master's thesis in Polymer Engineering. Université Ferhat Abbes Sétif 1, 2014.

CHAPTER I
HIGH-DENSITY POLYETHYLENE (HDPE)

I.1 Introduction

Polyethylene (PE) is a widely used polymer, and occupies a very dominant position in everyday life. It is generally used in all industrial sectors without exception, and more particularly in the field of drinking water and gas pipes. Knowledge of the various properties of this material is essential, and its worldwide consumption is increasing rapidly. In this first chapter, we present the definition of a PE, its chemical structure and its classification. Secondly, this chapter contains a detailed description of the main synthesis processes for high-density polyethylene (HDPE) and its properties.

I.2 The polyolefins

Polyolefins are materials derived from the polymerisation of olefins, i.e. hydrocarbon monomers with the general formula H C=CR$_{21}$ R$_2$. Or' R$_1$ and R$_2$ are groups such as: H; CH2; -CH2 -CH-(CH3)2... The main industrial polyolefins are : Polyethylenes (PE); polypropylenes (PP) and polyisobutylenes (P-IB) [1].

I.3 The Polyethylene

I.3.1 Definition

Polyethylene (PE) is the world's most synthetic plastic. Part of the polyolefin family, PE is a semi-crystalline thermoplastic material [2]. It is obtained by polymerising ethylene. Figure I.1 shows the ethylene polymerisation reaction.

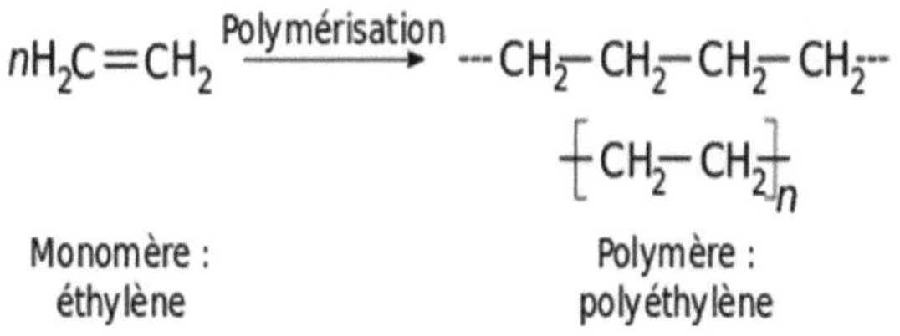

$$n\mathrm{H_2C}=\mathrm{CH_2} \xrightarrow{\text{Polymérisation}} \text{---CH}_2\text{--CH}_2\text{--CH}_2\text{--CH}_2\text{---}$$

$$\left[\text{CH}_2\text{--CH}_2\right]_n$$

Figure I.1: Polymerisation of polyethylene [2].

I.3.2 The polyethylene structure

The chemical structure of polyethylene is shown in Figure I.2 :

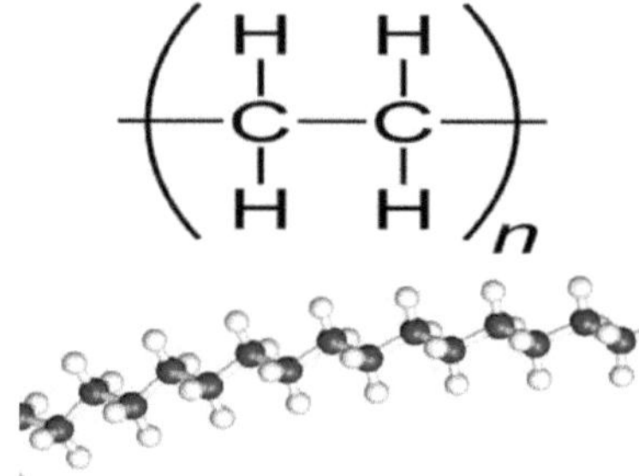

Figure I.2: Diagram showing the chemical structure of polyethylene [3].

I.3.3 Classification of polyethylenes

PEs are classified according to the type of polymerisation into three categories: Linear Low Density Polyethylene (LLDPE), Low Density High Pressure Polyethylene, and High Density Polyethylene (HDPE).

• Linear low density polyethylene (LLDPE)

Linear low-density polyethylenes are obtained by copolymerisation of ethylene and one or more olefins (butene-1, hexene-1, octene-1, tetramethyl-4- Pentene-1) under a pressure of less than 107 Pa, in the presence of Ziegler or Phillips type catalysts [4]. These products have characteristics similar to those of polyethylenes based on high-pressure densities (density varying from 0.9 to 0.94 and a melting point varying from 115-128°C).

• Low-density, high-pressure polyethylene

Low-density, high-pressure polyethylene is a polymer with long and short a branches manufactured by radical initiation using high-pressure processes with a density of 0.910 to 0.935 g/cm^3 [5].

• High-density polyethylene (HDPE)

High density polyethylene, also known as "low pressure" polyethylene, is obtained by polymerisation under less severe conditions than LDPE, with a polymerisation pressure of less than 50 bar and a temperature of around 100°C. HDPE has a crystallinity of 93% and a melting temperature of between 130°C and 145°C. HDPE chains are much more aligned than those of LDPE, which explains its high density (0.96 g/cm^3) [6].

I.3.4 High-density polyethylene (HDPE)

I.3.4.1 History

High-density polyethylene is produced by polymerising ethylene at low pressure, either alone or with cosmonomers. The first production units date back to the mid-1950s. The first was built in 1955 by Phillips in Texas. Hoechst then started up the first unit using the Ziegler process in 1956. In the 1960s, improvements were made to the Ziegler process with the use of super-active catalysts, eliminating the need for costly removal of catalytic residues. The most recent developments come from gas phase polymerisation processes
BASF built its first unit in 1964, Union Carbide definitively improved the gas-phase process and industrialised it in the 1980s, and today many licensees use this technology. Other gas-phase processes were subsequently developed and improved [7].

I.3.4.2 General Description (HDPE)

HDPE is a [8] :

► Semi-crystalline, whitish, semi-opaque commercial thermoplastic.
► The simplest and least expensive polymer.
► An environmentally-friendly material: its manufacture is clean, produces little waste and emits no harmful substances; it is 100% recyclable and requires very little processing.

energy.

I.3.5 Properties of High Density Polyethylene

HDPE is a family of polymeric materials with unique chemical, physical, mechanical, thermal, electrical and rheological properties.

I.3.5.1 Mechanical properties

At 23°C, HDPEs are above their glass transition temperature (Tg = -100°C) and their amorphous phase is rubbery, which affects their mechanical properties (**Table I.1**) [9].

Table I.1. Mechanical properties of high-density polyethylenes [9].

Properties	Units	HDPE
Density	g/cm3	2:0,955
Fluidity index (190° c)	g/10min	O,3-18
Stress at yield point (tensile)	MPa	25-30
Breaking strength	MPa	30-35
Breaking elongation	%	500-1100
Tensile modulus of elasticity	MPa	800-1100

I.3.5.2 Physical properties

High-density polyethylenes are opaque in thick layers and transparent in films. The increase in crystallinity results in a decrease in stability and diffusivity (**Table I.2**) [10].

Table I.2: Physical properties of HDPE [9].

Light transmission	Density g/cm3	Water absorption %	Crystallinity rate %
Wrong	0 ,95	0,01	70 à 80

I.3.5.3 Chemical properties

High-density polyethylenes are very chemically stable. At temperatures below 60°C, they are practically insoluble and are not attacked by acids (except oxidising agents), bases or salt solutions. They are insoluble in water, but in their natural state they are sensitive to the action of ultraviolet light in the presence of oxygen. filled with carbon powders, 2 to 3% or light stabilisers. They are sensitive to fire and to stress cracking in the presence of soap, alcohol, etc. [9].

I.3.5.4 Thermal properties

The melting point of HDPE is between 120 and 136°C. Thermal conductivity and the coefficient of linear thermal expansion are a function of the degree of crystallinity, and are higher for homopolymers than for copolymers [11].

I.3.5.5 Properties rheology

In its molten state, HDPE exhibits non-linear viscoelastic behaviour, meaning that its viscosity decreases with increasing shear. During extrusion, the polyethylene is subjected to a shear rate gradient from the extruder barrel to the die. It is therefore important to know the viscosity-shear rate curve over the entire shear range. Measurement of the melt flow index (MFI) provides an estimate of the viscosity at a given shear rate [11].

I.3.5.6 Electrical properties

HDPE has excellent electrical insulation properties regardless of its molecular weight and crystallinity. Its low relative permittivity and low dielectric dissipation factor make it a material of choice for electrical insulation [12].

I.3.6 HDPE applications

HDPE has a very wide range of applications, for example [13] :

❖ **Agriculture**
- Films ;
- Fishing nets ;
- Irrigation pipes ;
- Crates.

❖ **Packaging**
- Food (oil cans) ;
- Cosmetics ;
- Cleaning products.

❖ **Industry**
- Natural gas and water pipes ;
- Technical and automotive parts ;
- Containers.

I.3.7 Advantages and disadvantages of HDPE

HDPEs have a number of advantages, but they also have disadvantages [13].

a / Benefits

• Easy to install ;

• Excellent electrical insulation properties ;

• Impact resistance ;

• High chemical inertia ;

• Food quality ;

• Loss of the permeability of PEs, not only to water, but also to air and water vapour.

• Hydrocarbons.

b / Disadvantages

• Sensitivity to UV in the presence of oxygen ;

• Sensitivity to stress cracking ;

• Poor heat resistance ;

• Strong bonding.

I.3.8 Cycle de vie de PEHD

The production of HDPE pipes does not release any waste into the environment. Production waste is totally recycled on site, and the water used to cool the pipes produced circulates in a closed circuit, so there are no environmental discharges to worry about. Polyethylene is the most widely used material in the world for its reliability. It is an extremely hard-wearing material, which explains why it can be used in all climates and why water resources can be conserved to a considerable extent [14].

I.3.9 Recycling HDPE

High-density polyethylene is a 100% recyclable material that requires no specific reprocessing at the end of its life. It can be shredded and used in other applications. It can also be recovered by incineration with energy recovery [15].

I.3.10 Production unit (HDPE)

The high-density polyethylene (HDPE) production unit is located in the CP2K complex on the coast about 6 km east of the wilaya of Skikda, at an average height of about 6 metres above sea level [16]. This unit has a production capacity of 130,000 t/year in the Skikda industrial zone, and comprises a single production line [17].

The main raw materials used by the complex are :

- Ethylene from nearby CP1K or imported ethylene

- Isobutane from GL1K, also located nearby.

- hexene

- The catalyst

I.4 Conclusion

High-density polyethylene (HDPE) has been receiving double attention for many years in various applications because of its intrinsic properties. In this chapter, we present a general description of the treatment processes for this polymer, as well as its main characteristics. A presentation of the structure of the HDPE production unit in the wilaya of Skikda, Algeria, has also been developed.

References :

[1]J. P. TROTIGNE, J. VERDU, A. DOBRACZ and M. PIPERAND, "Plastics, Structures, propriétés, mise en œuvre, normalisation", Nathan, Paris, (1996), pp: 53-156.
[2]J. M. Berthelot, Matériaux composites comportement mécanique et analyse des structures, 4th edition, Lavoisier, 2005, pp. 1-63.

[3]"Polyethylene". http://fr.wikipedia.org/wiki/Polyéthylène.
[4]M.M, Jérôme. "Synthèse et caractérisation de silicates de calcium hydratés hybrides", PhD thesis, Université de Paris URF scientifique d'Orsay, Paris, 2003.

[5]S, Fûzesséry. Low density polyethylene, Techniques de l'ingénieur, A 3310,1-34.
[6]S.Sofiane, "contribution à étude expérimentale un polyéthylène à haute densité (PEHD)- effet de la température et de vitesse de déformation", magister thesis; Option: matériaux avances, université de chalef, 2006.
[7]H.Amina-L.Zaineb,<<Etude .l'effet de quelque charge végétales sur les propriétés du polyéthylène haute densité PEHD>>.mémoire de master 2 génie des polymers ,Skikda universities 2919-2020.
[8]B.Seddik,G.Farid<<Contribution à l'étude expérimentale d'un polyéthylène à haute densité (PEHD) -Effet de la température et de la vitesse de déformation>>mémoire de magister en matériaux avances, universités de Badji Mokhtar Annaba 2007.

[9]F. Montagne and G. W. Ehrenstein, Polymeric materials. Structure, properties and application, Hermès Science publication, 2000.

[10] Chiali Group Catalogue, Headquarters and General Management, Chiali Tubes / Société de Transformation de Plastique et Métaux, Zone Industrielle. Voie A. Bp 160 Sidi Bel Abbès, 22000Algérie.

[11] A.Hanane, B.Dounia "Influence of additives (carbon black) on the quality of high-density polyethylene", end-of-training professional project for engineering degree, IAP-Boumerdes school, 2017.
[12] M. Fontanille, P. Vairon ; Polymérisation ; Ed. Techniques de l'ingénieur, Traité plastiques et composites ; (A3 040).

[13] "Knowledge of polyethylene", ELF ATCHEM technical documentation, October 1995.

[14] S. Thomas, M.J. Jacob, B. Francis, K.T. Varughese, "Effect of chemical modification on properties of hybrid fiber bio-composites". Composites: Part A, 2008, Vol. 39 (2).

[15] C.Hacene. "Synthesis and characterization of HDPE/CaCO3 composites". Master's thesis in Chemical Analysis in Industrial Control and Environment, University of 20 August 1955 Skikda, 2011.

[16] Lyes BOUDIAF, Sontrach une compagnie pétrolier et gazière intégrée, Direction générale, page "8-10".

[17] H. Annab, A. Segueni "INFLUENCE OF ADDITIVES (SILICE) ON THE PROPERTIES OF HIGH DENSITY POLYETHYLENE (HDPE 5502)" SONATRACH CP2K SKIKDA", master's thesis, Université de Larbi BenM'hidi Oum El Bouaghi, 2019.

CHAPTER II
GENERAL INFORMATION ON COMPOSITE MATERIALS

II.1 Introduction

The study and design of composite materials has attracted a great deal of interest in various areas of modern chemistry for almost a century. Their performance is often far superior to that of homogeneous materials, which opens up very promising prospects for their use [1]. This part of the manuscript presents a literature review covering general information on composite materials and their classifications.

II.2 Definition

In a very broad sense, the word "composite" means "made up of two or more different parts". A composite is also defined as a material that combines at least two components that are not miscible but have a high adhesion capacity. The combination of these elements results in a material whose properties are superior to those of the elements taken separately [2,3]. Composite materials are not new; they have been used by man since ancient times, such as wood and cob, which are materials used in everyday life [4].

II.3 Constituents of composite materials

In general, the main constituents of a composite material are :

► The matrix.
► Reinforcement.
► Fillers and additives.

Figure II.1 shows the constituents of a composite material.

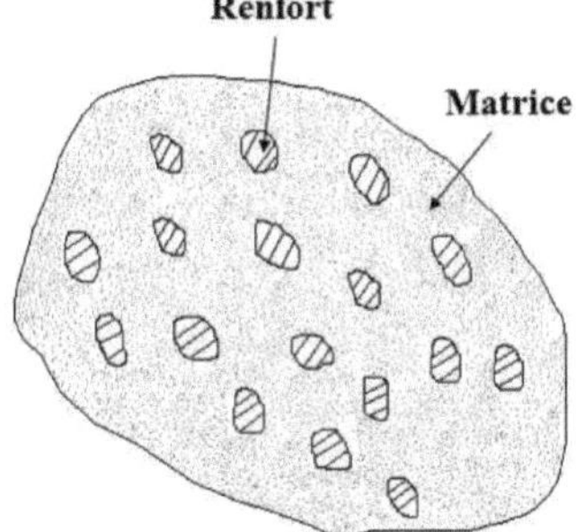

Figure II.1: The constituents of a composite material [5].

II.3.1 The matrix

The matrix is the element that binds and holds the fibres together. It distributes stress (resistance to compression or bending) and provides chemical protection for the fibres. A classification of the types of matrix commonly encountered is given in Figure II.2 [5].

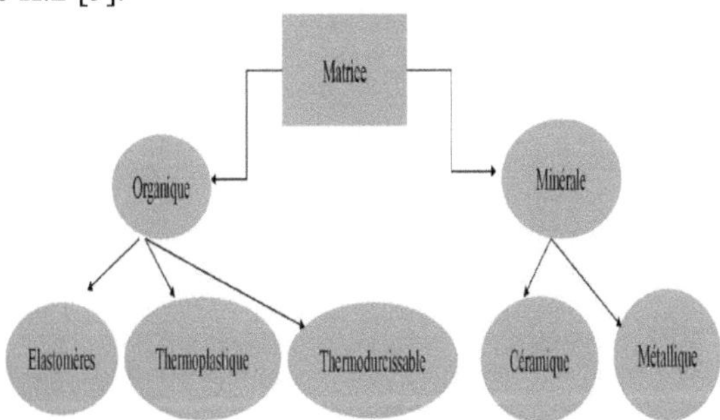

Figure II.2: Flow chart showing the different types of matrix.

II.3.2 Reinforcement

It is the main constituent carrier in the composite (shape, volume). It gives composites their mechanical characteristics: rigidity, resistance to breakage and hardness. Reinforcements can be mineral (glass, boron, ceramic, etc.) or organic (polyester or aramid). Glass and carbon fibres are the most commonly used [5].

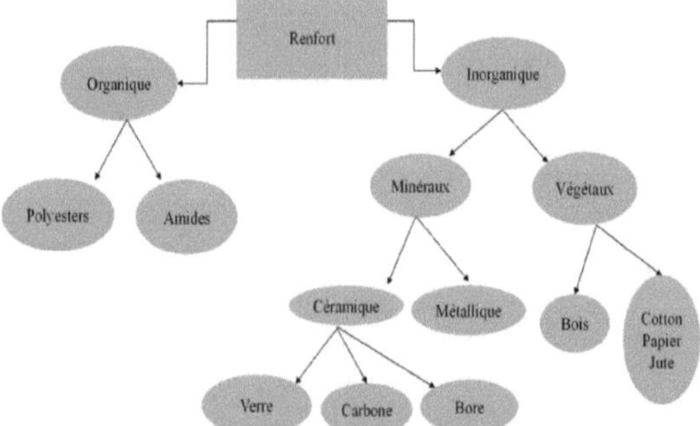

Figure II.3: Flow chart showing the classification of reinforcements according to their origin [5].

II.3.3 The aim of incorporating an additive into a polymer

The intrinsic properties of an object (micrometric or nanometric) can be used to give the material specific properties such as magnetic or electrical properties and/or to modify its thermal, mechanical or optical properties [6].

II.4 Classification of composite materials

Composites can be classified according to the shape of the components or their nature [7].

II.4.1 Classification by constituent form

Depending on the shape of the components, composites are classified into two broad classes:
· Fibre composite materials.
· Particle composite materials.

II.4.1.1 Fibre composites

A composite material is a fibre composite if the reinforcement is in the form of fibres. The fibres used can be in the form of continuous fibres, or in the form of discontinuous fibres, short fibres, chopped fibres, etc. The arrangement of the fibres and their orientation make it possible to modulate the mechanical properties of the composite materials, to obtain materials ranging from highly anisotropic to anisotropic in a plane. The importance of fibre composites justifies an exhaustive study of their mechanical behaviour [7].

II.4.1.2 Particle composites

A composite material is said to be particle-based when the reinforcement is in the form of particles. A particle, as opposed to a fibre, does not have a preferred dimension. Particles are generally used to improve certain material properties such as stiffness, temperature resistance, abrasion resistance, etc. The choice of matrix/particle combination depends on the desired properties of the composite [7, 8].

II.4.2 Classification according to the nature of the constituents

Depending on the nature of the matrix, composite materials are classified as polymer matrix, metal matrix or mineral matrix composites. Various reinforcements are associated with these matrices [9, 10].

II.4.2.1 Polymer matrix composites

Polymers are characterised by low density, relatively low mechanical strength and high deformation at break. The main advantages of these composites are the relatively mature manufacturing process and low weight. This type of composite has been developed mainly for aeronautical applications where weight reduction is essential.

II.4.2.2 Metal matrix composites

In these composites, metallic materials such as aluminium and titanium are reinforced by generally non-metallic reinforcements, often ceramics. By the very nature of the composite, metal matrix composites have better mechanical properties or are more adaptable to loading than their monolithic matrices. Their applications in car engines are well established.

II.4.2.3 Mineral matrix composites (ceramics)

Ceramic matrices such as glass and silicon carbide (SiC) can be combined with reinforcements such as metals, carbon and ceramics. Their development is aimed at improving mechanical properties such as toughness and thermal shock resistance of monolithic ceramics. These composites are used in harsh environments, such as rocket engines, heat shields and gas turbines [11]. There are also two types of composite for different markets [12]:

- Mass-produced" composite materials, which have weaker mechanical properties but a cost compatible with mass production.
- high performance" composite materials, with high specific mechanical properties and a high unit cost. These are the materials most commonly used in the aerospace industry.

II.5 Structure of composite materials

Composite material structures can be classified into three types [13] :
• Monolayers.

• Laminates.

• Sandwiches

II.5.1 Monolayer

Monolayers are the basic elements of composite structures. Unidirectional fibres placed in the median plane are embedded in a polymer matrix. They are characterised by the type of reinforcement used: long fibres (unidirectional or not), short fibres, in the form of fabrics or ribbons [14].

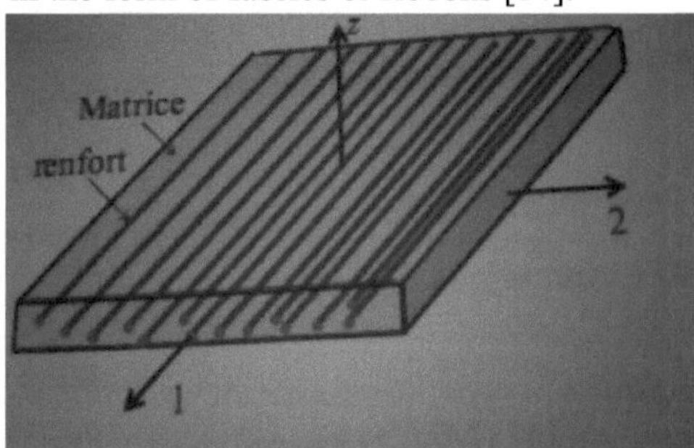

Figure II.4: Composite layer.

II.5.2 Laminate

A laminate is made up of a stack of monolayers, each of which has its own orientation relative to a reference frame common to all the layers and referred to as the laminate reference frame. The choice of stacking, and more specifically the orientations, will give the laminate specific mechanical properties [13].

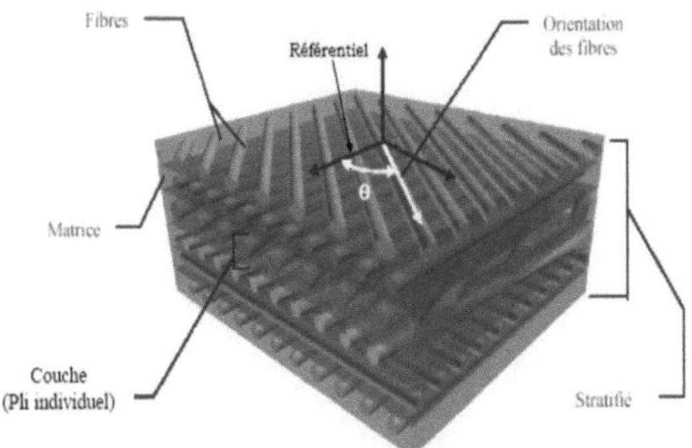

Figure II.5 : Composite laminates [13].

II.5.3. Sandwiches

Composite structures subjected to flexural or torsional stresses are generally made of sandwich materials. A sandwich structure consists of a core and two skins made of composite materials. They are bonded together using a resin that is compatible with the materials involved. The most commonly used cores are honeycomb, corrugated or foam. The skins are generally made of laminated structures [13].

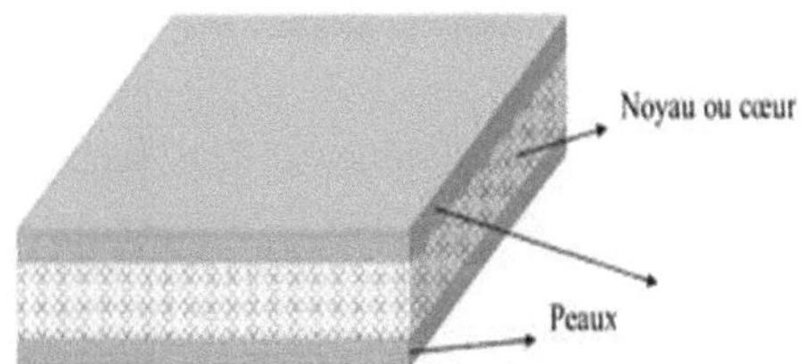

Figure II.6: Structure of sandwich composites

II.6. Advantages and disadvantages of composite materials

II.6.1 The advantages of composite materials :

Composites are preferred to other materials because they offer advantages in terms of :
*Their lightness
*Their resistance to corrosion and fatigue.

*They are insensitive to products such as greases, hydraulic fluids, paints and solvents.

* Their ability t o take several forms, incorporate accessories and enable the noise reduction [14].

II.6.2   The disadvantages of composite materials

However, there are a number of drawbacks to their spread:
* The costs of raw materials and manufacturing processes.
*Waste management and increasingly strict regulations [14].

II.7. Applications for composite materials

These days, composite materials are playing a major role in a number of fields:

■ Electricity and electronics,
■ Building and public works,
■ Road, rail, sea, air and space (especially military) transport,
■ Health (medical instrumentation) [15].

II.8 Conclusion :

Composite materials are very varied, available everywhere, and are currently experiencing a major boom in all areas of application. Their characteristics depend directly on the choice of constituents. This section defines composite materials in a general way, and in particular their types, properties and areas of application.

References

[1] D. Roy, S. Massey, A. Adnot and A. Rjeb, Action of water in the degradation of lowdensity polyethylene studied by X-ray photoelectron spectroscopy, Express Polym. Lett, Vol. 1, 2007,pp. 506-511.

[2] J. M. Berthelot, Matériaux composites comportement mécanique et analyse des structures, 4th edition, Lavoisier, 2005, pp. 1-63.

[3] S. Étienne, D. Laurent, E. Gaudry, P. Lagrange, J. Le dieu and J. Steinmetz, Les matériaux de A à Z, Dunod, 2008, pp. 54-63.

[4] D. Gay, Composite Materials, 5th edition, Lavoisier, Paris, 2005, pp. 19-260.

[5] B. Bouchra "Characterisation of a glass fibre/epoxy laminated composite material in static 3-point bending". Master's thesis in Mechanical Engineering, Université Badji Mokhtar Annaba, 2018.

[6] B. Sabrina and LOUCHI Soumaya. "L'effet de l'incorporation d'une charge végétale (Cyprès) sur Les propriétés de PEHD". Master's thesis in Polymer Engineering, University 20 August 1955 Skikda, 2019.

[7] B. Ringuette. Composite materials based on hemp fibres. Master's thesis. Université Laval- Québec, 2011.

[8] J.M. Berthelot. Composite materials: Comportement mécanique et analyse des structures. 5th ed, Technique et documentation, Lavoisier : Paris, 2005.

[9] M. Geier, D. Duedal. Guide pratique des matériaux composites. Technique et documentation, Lavoisier: Paris, 1985.

[10] W. Kurz, J. P. Mercier, G. Zambelli et al, Traité des matériaux : Introduction à la science des matériaux. Presses Polytechniques et Universitaires Romandes: Paris, 1995.

[11] Z. Gürdal, R.T. Haftka, P. Hajela. Design and optimization of laminated composite materials. Wiley-Interscience Publication : Canada, 1999.

[12] M. Reyne. Traité Plastiques et Composites, Paris: Techniques de l'Ingénieur: Paris, AM 5002, 1998, p. 1-6.

[13] M. Haddadi. Numerical study with experimental comparison of the thermoplastic properties of polymer matrix composite materials. Master's thesis. Université Al Hadj Lakhdar-Batna, 2011.

[14] K.Abd raouf. "Effect of elaboration parameters on the mechanical

behaviour of a bio-composite". Master's thesis in mechanical engineering. Université Mohamed Boudiaf M'sila, 2016.

[15] : B. Bouchra. "Characterisation of a Glass fibre/epoxy laminated composite material in static 3-point bending". Master's thesis in Mechanical Engineering, Université Badji Mokhtar Annaba, 2018.

CHAPTER III
GENERAL INFORMATION ON PLANT FIBRES

III.1 Introduction

Over the last few decades, natural fibres have received increasing attention both from the academic world and from various industries in the different fields of composites. They are widely used because they are readily available, so their use enhances the local resources of a given country, while respecting the environment [1]. This chapter provides an overview of natural fibres, in particular plant fibres, their classification, structure and chemical composition, properties and applications.

III.2 Definition of natural fibres

Natural fibres are produced from plants, animals or minerals and are not chemically transformed. It can be said that, apart from the mechanical transformations carried out on the raw material to convert it into fibre, there has been no modification [2].

III.3 Classification of natural fibres

Most natural fibres are of plant, animal or mineral origin.
Plant-based: extracted from plants, fruit and trees such as cotton, linen, jute and hemp, etc.

Animal: extracted from the hair of animals such as sheep, goats, llamas, etc.
Mineral: minerals with a fibrous texture are found in nature and are toxic materials, such as asbestos [3].

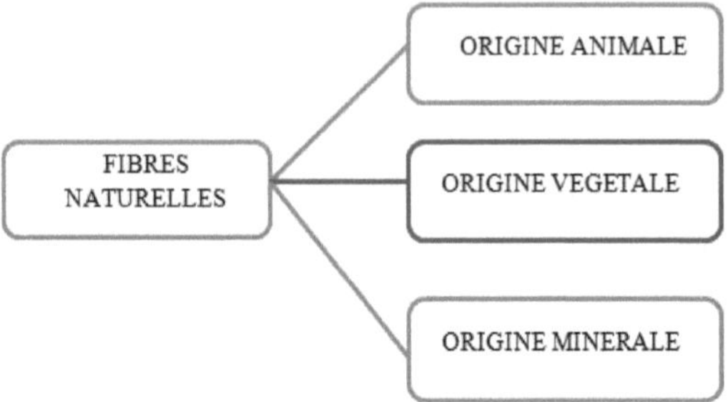

Figure III.1: Classification of natural fibres.

III.4 Plant fibres :

III.4.1 Definition of a plant fibre

Plant fibre is a fibrillar biological structure composed of cellulose, hemicellulose and lignin, with a relatively small proportion of non-nitrogenous extractable matter, crude protein, lipid and mineral matter. The ratio of these constituents depends largely on the species, age and organs of the plant **[4]**.

III.4.2 Classification of plant fibres

Plant fibres are divided into four groups according to their origin: leaf, stem, wood and surface fibres **[5]**. **Figure III.2** shows the main classes of plant fibres.
Leaf fibres: these fibres are obtained by rejecting monocotyledonous plants. They are made by overlapping the bundles that surround the leaves to strengthen them. These fibres are hard and rigid. The most widely grown types of leaf fibre are sisal and abaca **[6]**.
Stem fibres: stem fibres come from the stems of dicotyledonous plants. Their function is to give good rigidity to plant stems. They are marketed in the form of horn bundles and in any length. The most commonly used stem fibres are jute, flax, ramie, kenaf and hemp **[7]**.
Wood fibre: Wood fibre comes from the shredding of trees such as bamboo or reeds. They are generally short **[6; 8]**.

Surface fibres: Surface fibres generally surround the surface of the stem, fruit or grain. Grain surface fibres are the largest group in this family of fibres. Examples include cotton and coconut (coco) **[9]**.
Sap fibres: such as latex or rubber (rubber tree) **[10]**.

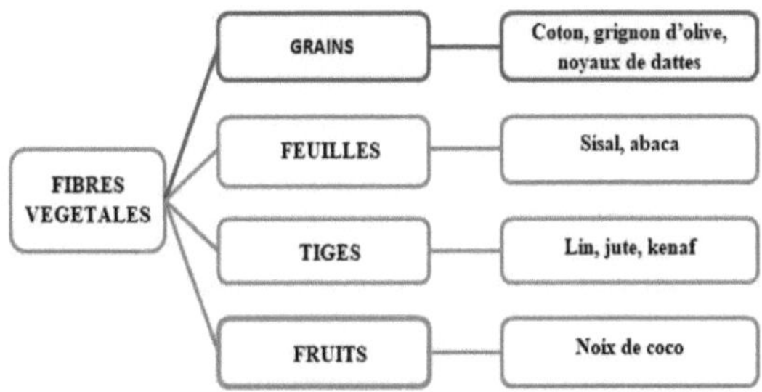

Figure III.2: Classification of plant fibres.

III.4.3 The structure of a plant fibre

Plant fibre can be likened to a composite material whose reinforcement is provided by cellulose fibrils embedded in a matrix formed of hemicellulose and lignin, which is a very rigid structure **[11]**. The fibrils are arranged in a helix and form an angle with the fibre axis known as the "microfibril angle".

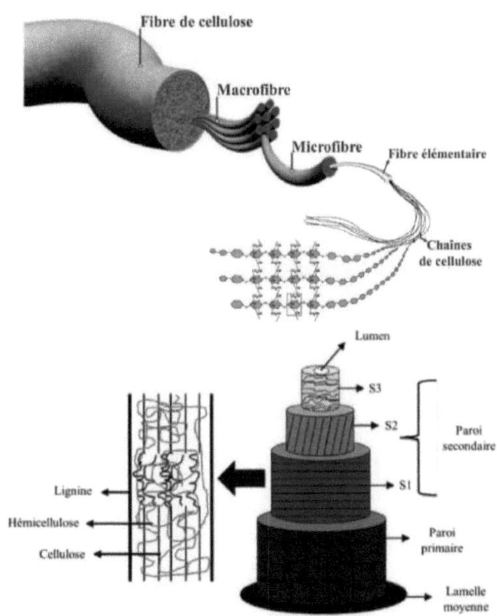

Figure III.3: The structure of a plant fibre.

III.4.4 The chemical composition of a plant fibre

Plant biomass is made up of several macromolecules that are closely linked to each other within the plant wall. In the case of plant fibre stems, there are three main compounds in their walls: cellulose, hemicelluloses and lignins. Lignin acts as a matrix coating cellulose, which is a very rigid structure [1], and hemicelluloses act as an accounting agent at the interface between these two elements [12]. These three main constituents are supplemented by low molecular weight substances, extractables and ash [12, 13].

A. Cellulose

It is the main component of plant fibres. It is a leading natural polymer. In general, plant fibres are made up of a chain of cellulose fibres [14].

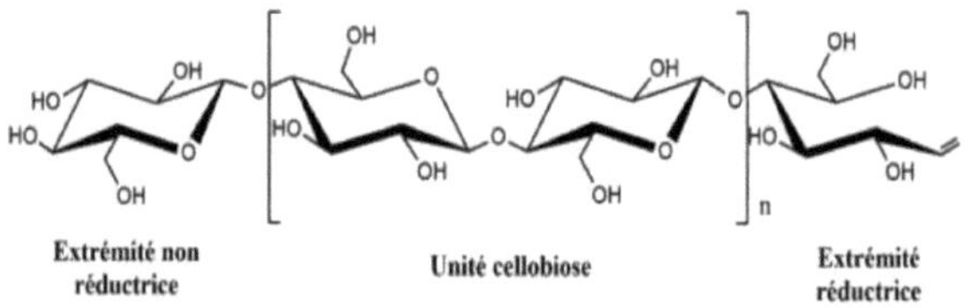

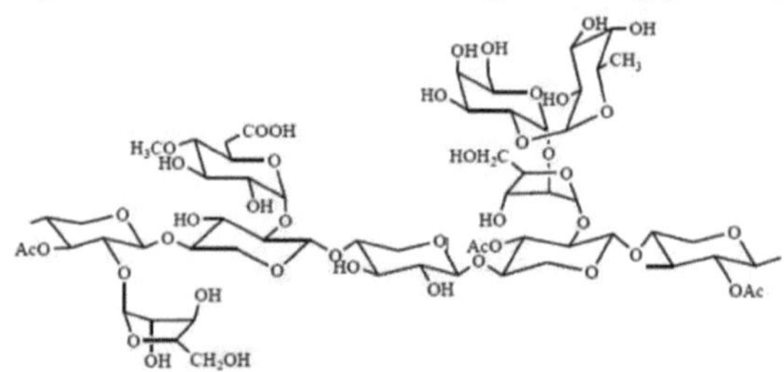

Figure III.4: Representation of the cellulose chain.

B. Hemicellulose

The hemicellulose present in all the walls of these fibres is a short-chain polysaccharide, branched and folded on itself. It is the component responsible for fibre elasticity and allows the walls to elongate during growth [14].

Figure III.5: Molecular structure of hemicellulose [15].

C. Lignin

Lignin is the glue that binds plant fibres together and their walls. It is a three-dimensional polymer derived from the copolymerisation of three phenylpropenoic alcohols **[14]**.

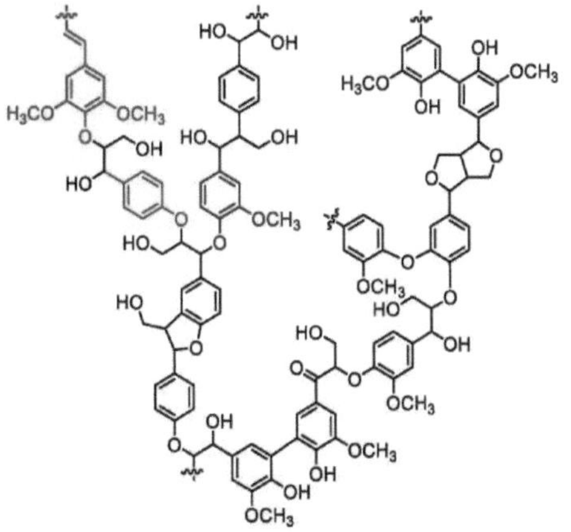

Figure III.6: Molecular structure of lignin.

III.4.5 Physical and mechanical properties of plant fibres

The mechanical properties of plant fibres are determined by the intrinsic characteristics of these fibres (chemical composition, cellulose, hemicellulose, lignin and pectins, fibre structure, cross-section, porosity, micro-fibril angle, form factor, length/diameter ratio, etc.), by anthropogenic characteristics, or by independent and variable characteristics (moisture content, location of fibres in the stem, natural defects, growth conditions, etc.). Mechanical properties such as Young's modulus (GPa), stress at break (MPa), elongation at break (%), and sources are influenced by many factors, including plant origin, variety, growth and harvesting **[17]**.Plant fibres are physically characterised by their length, diameter and density.

Table III.1: Physico-mechanical properties of several natural fibres **[18]**.

Fibre	Tensile strength(MPa)	Module d'Young(GPa)	Elongation at break % (mm)	Density (g/cm)3
Abaca	400	12	3-10	1.5
Linen	345-1035	27.6	2.7-3.2	1.5
Hemp	690	70	1.6	1.48
Jute	393-773	26.5	1.5-1.8	1.3
Kénaf	930	35	1.6	-
Sisal	511-635	9.4-22	2.0-2.5	1.5
Ramie	560	24.5	2.5	1.5
Pineapple	400-627	1.44	14.5	0.8-1.6
Coco	175	4-6	30	1.2

III.4.6 Advantages and disadvantages of plant fibres.

Some of the advantages and disadvantages of plant fibres are **listed in Table III.2.**

Table III.2: Advantages and disadvantages of plant fibres **[19]**.

BENEFITS	DISADVANTAGES
-low cost. -Biodegradability. CO2-neutral. -No skin irritation when exposed to fibre handling. -No residues after incineration. -Renewable resource. -Requires little energy to operate produced. -Highly specific mechanical properties (strength and rigidity). -Good thermal and acoustic insulation	-Water absorption. -Low dimensional stability. -Low thermal resistance (200 to 230°C max) -Anisotropic fibres. -Variation in quality depending on growing location, weathero....etc. -For industrial applications on request stock management. -Discontinuous reinforcement.

III.5 The plant filler used (spiny cactus leaf powder)

III.5.1 The spiny cactus

III.5.1.1 Origin

The original habitat of this plant dates back to Mexico, and today it is widespread in many countries around the Mediterranean basin. Scientists count over 1600 varieties of cactus, and the cactus plant can withstand drought and survive with less than 50 millimetres of rain a year.

Figure III.7 : The spiny cactus.

III.5.1.2 Definition

The thorny cactus is an oily plant with a medium height stem rich in juices. between 1.5 and 3 metres, in the shape of a pallet or column. The fruit and leaves of the spiny cactus are rich in fibre, vitamins, minerals and antioxidants. It also has anti-cholesterol and anti-inflammatory properties.

Figure III.8: The leaf of the spiny cactus.

III.6. Conclusion

Natural fibres are now becoming an attractive alternative to synthetic fibres because of their indefinite value (per unit weight). They have been used to develop various composites based on thermosetting and thermoplastic matrices. As with any composite, its mechanical, thermal and physical properties depend on the properties of the matrix and reinforcement, as well as the fibre loading, fibre source and process.

References :

[1] A. Y. Nenonene. Elaboration and mechanical characterisation of kenaf stem particle boards and bio-adhesives based on bone glue, tannin or mucilage. PhD thesis. Toulouse - Jean Jaures University, 2009.

[2] https://youmiwi.com/blogs/youmiwi/une-fibre-naturelle-c-est-quoi.

[3] Omiri imen Yamina, "The effect of natural fibre treatment on the damage to an polymer concrete", master thesis, msila university ,2014/2015.

[4] F. Laurans, A. Déjardin, J. Pilate. Physiology of wood fibre wall formation.

Adv. Compos. Mater. 2006, 16, p. 27-39.

[5] Mostar Abdessamed, "influence des ajouts de fins minérales sur les performances mécaniques des bétons renforcé de fibre végétales de palmier dattier", Kasdi Merbah University, Ouargla, civil engineering option, November 2006.

[6] SWAMY, R. H. S, AHUJA, B. M, KRISHAMOORTHY, S? Behavior concrete reinforced with jute, coir, bamboo fibres. The international journal of cement composite and light weight concrete, volume 5, p 13 N°1, 1984.

[7] COUTTS, R.S.P., Flax fibers as a reinforcement in cement mortar , the International journal of cement composites and lightweight concrete, vol.5 N°4, pp 257-262, 1983.

[8] KRIKER. A. Caractérisation des fibres de palmier dattier et propriétés des bétons et mortiers renforcés par ces fibres en climat chaud et sec. PhD thesis, ENP, Algiers, 2005

[9] BLEDZKI, A. K and GASSAN. J? Composites reinforced with cellulose based fibers ELSEVIER, Progress in Polymer science, volume 24, pp.221-274, 1999.

[10] https://textileaddict.me/les-fibres-naturelles/

[11] Brosse N. **"Fibres, fibrous materials and eco-materials"**. The Canadian

Conference on Industrial Bioproduct Innovation (BOIP), Power Point presentation, (2008).

[12] E. Bodros, C. Baley. Etude des propriétés de biopolymer renforcés par des fibres de lin aléatoirement dispersées dans le plan de stratification", properties at interfaces and composites. Université de Bretagne-Sud, 2006.

[13] M. Nardin. Fibre-matrix interface in composite materials - application to plant fibres. Adv. Compos. Mater, 16, p. 49-61, 2006.

[14] **Mokhtari Abdessamed** (Influence des ajouts de fines minérales sur les Performances Mécaniques des Bétons Renforcés de Fibres Végétales de Palmier dattier) magister thesis, University of Kasdi Merbah Ouargla, 2006.

[15] M.saiful Islam. The influence of Fibre Processing and Treatments on Hemp Fibre/Epoxy and Hemp Fibre/PLA Composites. PhD Thesis-University of Waikato, Hamilton, New Zealand, 2008.

[16] Monot, Claire. Contribution to the study of lignin-carbohydrate complexes (LCC) in wood. Study of the impact of the various stages of a sulphur-free bio-refinery process on LCCs. S.l. : univérsité Grenoble alpes, 2015.

[17] Martin N.A.M, Contribution à l'étude de paramètres influençant les propriétés mécaniques de fibres élémentaires de lin : Correlation avec les propriétés de matériaux composites, phd Thesis, Université de Bretagne Sud, 2014.

[18] O. Faruk, A.K. Bledzki, H.P. Fink, M. Sain. **"Bio composites reinforced with natural fibers".** Progress in Polymer Science, 37, 1552-1596, (2012).

[19] Talal, B. Use of a non-contact optical method to describe the mechanical behaviour of Wood/Plastic composites 'WPC '. PhD thesis from the University of Pau and the Pays de l'Adour, 2011

CHAPTER IV
MATERIALS AND METHODS

IV.1 Introduction

In this section, we focus primarily on the presentation of the different materials used in this study. On the other hand, the raw material preparation and treatment process, as well as the processing protocol used in the manufacture of the composites, are presented. This chapter also includes the main methods used to characterise the composites prepared.

IV.2 Materials used

IV.2.1 Resin

The polyethylene used in this study is HDPE reference 5502 produced by the POLYMED CP2/K unit in the Skikda industrial zone. This HDPE is obtained by catalysed polymerisation based on chromium oxide, according to the Philips process using low pressures. Its main characteristics are **shown in Table IV.1** :

Table IV.1: Main characteristics of HDPE 5502 **[1].**

Features	Methods	Values
Fluidity index (2.16Kg/190°C)	ASTM D1238	0.35g/10min
Density (23°C)	ASTM D1505	0.955g/10min
Hardness, shore D	ASTM D2240	67
Tensile strength at Rupture (50mm/min)	ASTM D 638	28 MPa
Elongation at break (50mm/min)	ASTM D 638	>600%
Flexural modulus	ASTM D 790	1200 MPa
Melting temperature	-	194 to 216°C

IV.2.2 The load used

PFCE spiny cactus leaf powder is the filler used in the preparation of the composites, and was recovered by grinding dried spiny cactus leaves.

Figure IV.1: The spiny cactus leaf.

IV.2.2.1 Preparing spiny cactus leaf powder:

The filler powder used was prepared from spiny cactus leaves in the following steps:Firstly, all the spines on the cactus leaves were removed with a brush, and then, after extraction of the gel found inside, the prickly cactus leaves were cut into small pieces and washed several times with distilled water. The leaves were then dried under solar irradiation until all traces of water were removed. The leaf pieces were recovered and ground using a small mortar until a fine powder was obtained. Finally, the ground powder obtained was passed through a flour sieve with a diameter of < 180µm. All the steps are shown in **Figure IV.2.**

Figure IV.2: Stages in the preparation of PFCE powder.

IV.2.3 Preparation of composite materials :

In the present study, several polyethylene-based formulations were prepared with levels of prickly cactus leaf powder varying from 0%, 5%, 15%, 25%. The quantities of HDPE and PFCE required are shown in **Table IV.2**.

Table IV.2: Composition of the different formulations.

Formulations(%) (HDPE/PFCE)	HDPE(g)	PFCE(g)
F0 (100/0)	150	0
F1 (95/5)	142.5	7.5
F2 (85/15)	127.5	22.5
F3 (75/25)	112.5	37.5

Firstly, and as a first step, each formulation of HDPE/PFCE composites is placed in a sealed bottle and mixed until the mixture becomes homogeneous.

Figure IV.3: The mixing bottle.

IV.2.3.1 Mixing

The composites were prepared in a two-cylinder mixer (calender) made by IQAP LAP at the CP2K Skikda unit, with a rotation speed of 32 rpm. The temperature was set at 170°C, with a mixing time of 10 min. The 1mm-thick composite sheets that will be used to prepare the specimens using the compression moulding process were recovered.

Figure IV.4: Mixing HDPE with PFCE on the mixer.

IV.2.3.2. Grinding

The leaves obtained were cut into small strips.

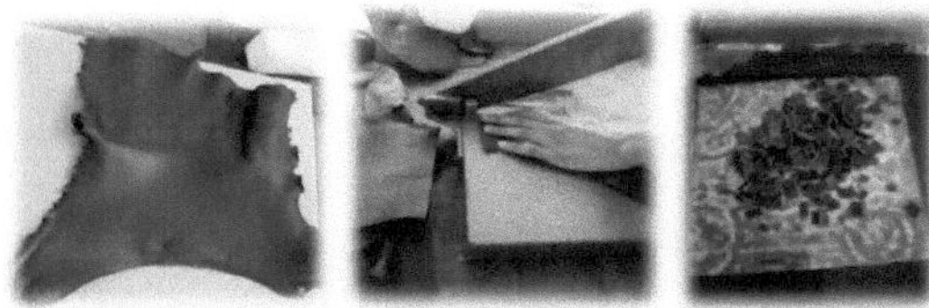

Figure IV.5: The leaf cut into small pieces (original photo) [3].

IV.2.3.2. Compression moulding

The films obtained by calendering are cut into small pieces and then inserted between two sheets of insulating Teflon sandwiched between two metal plates on a CARVER press. They were then heated to 190°C for a residence time of 15 minutes. Different shapes of 3mm-thick samples were obtained, which will be the subject of different characteristics.

Figure IV.6: The hydraulic press.

Figure IV.7: Preparation of test specimens.

IV.3 Characterisation techniques

This section describes the various techniques used for the physical and mechanical characterisation of the materials produced.

IV.3.1 Physical characterisation

IV.3.1.1 Density

The density at 23° is measured using the gradient density column technique using CEAST type 6001 equipment in accordance with ASTM D-1505. The samples to be analysed can be cut in any shape and must have dimensions that allow for the most accurate position. Care must be taken to cut the samples from the centre of gravity of a circle filled to a thickness of 2 mm, which has already been moulded using a hydraulic press. The samples, rinsed with isopropanol, are then introduced into the column, and their heights are plotted so that we can determine the average density values over three tests **[3]**. The density values for each sample were calculated using the following formula:

Density (at 23°C) = (Y/Z)*(B-A) +A(IV.1)

Or :
Y: distance between the sample and the low density float.
Z: distance between the two floats.
A: density of the upper 1er float.
B: density of 2eme lower float.

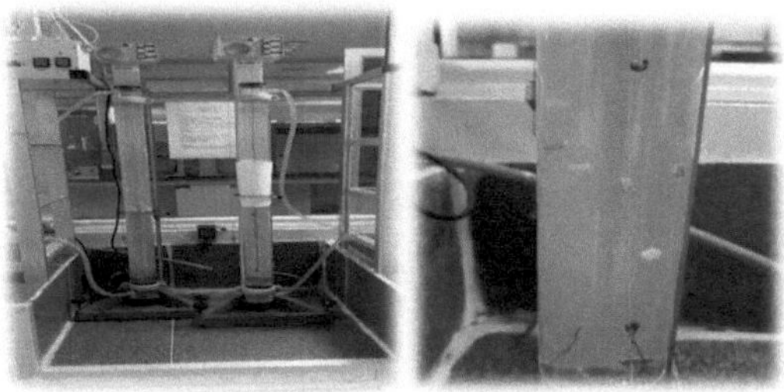

Figure IV.8: Principle of the density test [3].

IV.3.1.2   Water absorption test

The samples are dried in a vacuum oven at 100°C for 4 hours, then immersed in a beaker filled with distilled water at room temperature. Periodic samples are taken every 24 hours, the water on the surface of the sample is absorbed with absorbent paper and the sample is weighed again **[3]**.The dried mass of the sample after soaking in water is defined by the following relationship :

Water absorption rate (%) = (Mt- M0) / M0 ×100   (IV.2)

M0: dry sample mass

Mt: mass of sample after absorption

IV.3.2 Mechanical characterisation

IV.3.2.1 Tensile test

Tensile testing is commonly used to determine the mechanical behaviour of materials. Firstly, a tensile force is exerted on a cylinder of a given length, as the test results are influenced by the accuracy of the measurement of the dimensions of the specimen, until it breaks. Then, by recording the force F applied to the specimen by the traction machine and its progressive elongation. Tensile tests are carried out using a universal tensile testing machine of the type TesT GMBH.
- The test is carried out at room temperature (laboratory).
-The deformation speed is set at 20mm/min.
This test can be used to determine a number of standardised quantities, including:
a) **Modulus of elasticity:** This is the ratio of the maximum tensile stress a material can withstand to the deformation corresponding to its rigidity [4]. It is expressed as follows:

$$E = a / t : (IV.3)$$

With :
E: modulus of elasticity
a: stress (N/m²)
r: elongation or deformation (%)

b) Stress at failure: The tensile load borne by the specimen at the moment of failure per unit area, given by the following expression **[4]**.
$$a = F / S (IV.4)$$
F : tensile load supported by the specimen (N).

S: initial cross-section (m²).

c) Elongation at break: or deformation at break, is the change in length compared with the initial length **[4]**, given as a percentage.
$$t: = (AL / L0)*100(\%) \ (IV.5)$$

L0: initial length of the specimen.
L: final length of the test piece.

With :

$$AL = L - L0 \ (IV.6)$$

d) The elastic limit: is thefirst load value on the stress-strain curve, at which the curve's inclination becomes oriental.

Figure IV.9: The tensile test apparatus.

IV.3.3 Morphological characterisation by optical microscopy

In order to study the morphology of the materials and check the dispersion of the fibres in the composite materials, optical microscopy is used to take photos of the surface of the films obtained. The equipment used is an OPTIKA Microscopes ITALY optical microscope.

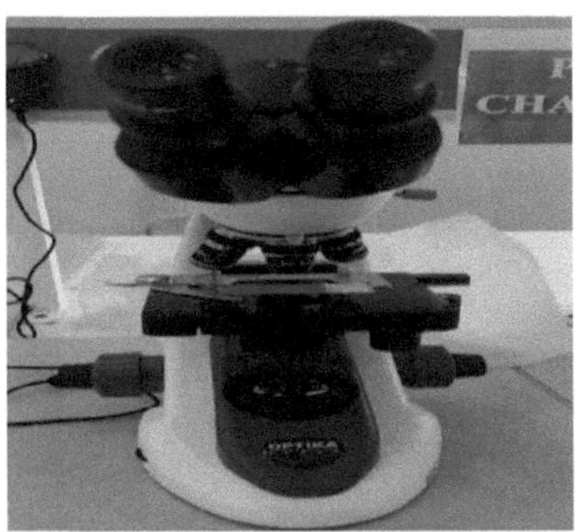

Figure IV.10: Optical microscope equipment.

References :

[1] Atamnia Amira and Benzedira Hadjer, (Etudes et valorisation d'une fibre biodégradable dans le domaine des composites), master's thesis in petrochemistry and polymer processes, université 20 Aout 1955 Skikda, 2017.

[2] MERAREB Imen and LEKHCHINE Souheyr. "Elaboration and characterization of a composite material with vegetable filler (walnut shell powder)". Master's thesis in Polymer Engineering, Université 20 aout 1955 SKIKDA 2019.

[3]BOUREGA Sabrina and LOUCHI Soumaya. "The effect of incorporating a vegetable filler (Cypress) on the properties of HDPE". Master's thesis in Polymer Engineering, Université 20 aout 1955 Skikda, 2019.
[4] Test method, www.ulttc.com/fr/prestations/methodes-d'essai/mécanique/shore-a-et- shore-d.html .

CHAPTER V
RESULTS AND DISCUSSIONS

V.1 Introduction

This part of the project covers the main results obtained. The mechanical, physical and morphological characteristics of HDPE/PFCE composites prepared with different filler ratios are discussed.

V.2 Physical study

IV.2.1 Density

The physical properties of the synthesised samples are studied by measuring the density of each of them. The values obtained are shown in the table below:

Table V.1: Density values for HDPE/PFCE composites.

Load rate (%)	0%	5%	15%	25%
Density	0.9541	0.9418	0.9326	< 0.93

The results show that the density of the composite materials is significantly lower than that of the pure matrix. The addition of a vegetable organic phase (light) to the polymer leads to a reduction in density. The decrease in density as a function of the increase in charge rate can be explained by the occupation of the free volume by less dense particles.

V.2.2 Water absorption test

In order to assess the rate of water absorption of the composite materials prepared, tests were carried out on the following materials have been applied. The test results are shown in **Figure V.1.**

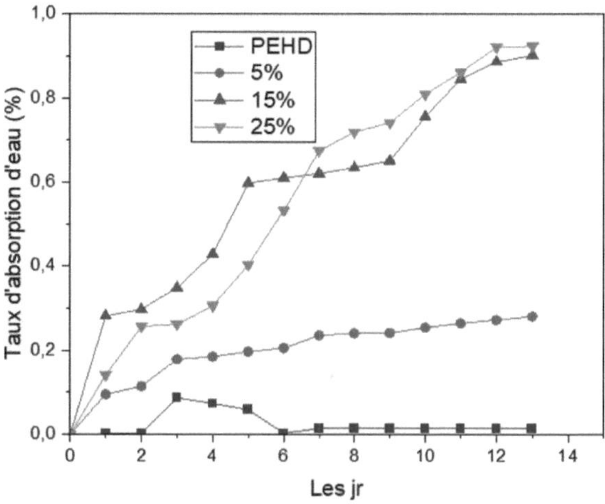

Figure V.1: Evolution of the water absorption rate of pure HDPE and composites HDPE/PFCE.

The results show that the water absorption rate of composites depends on directly from experience time. The rate of water absorption increases as the percentage of the load increases in the composite. A very high rate of water absorption is observed in the HDPE/PFCE composite material with 25% of the plant fibre. In the case of pure HDPE, very little water was absorbed compared with HDPE/PFCE composites filled with cactus powder, confirming the hydrophilic nature of the filler.

V.3 Study of the effect of loading rates on the mechanical properties of HDPE/PFCE composites

V.3.1 Mechanical properties of composites obtained by tensile testing

a) Maximum stress (yield point)

The elastic limit is the stress at which a material ceases to deform in a way that causes it to break. elastic, reversible and thus begins to deform irreversibly.

Figure V.2 shows the variation of the maximum stress as a function of the PFCE load rate.

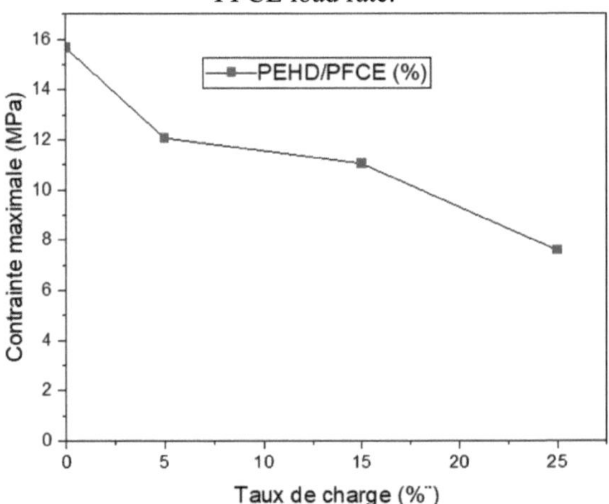

Figure V.2: Variation of the maximum stress in the elastic domain as a function of the PFCE load rate.

The data presented in this figure show a decrease in the maximum stress as a function of the increase in the percentage of PFCE. It can be said that the blends have lost their elasticity and gained rigidity through the introduction of the rigid PFCE particles; this decrease is also due to the fact that the PFCE filler is incompatible with HDPE.

b) Stress at break

The results of the stress at break are shown in **Figure V.3.**

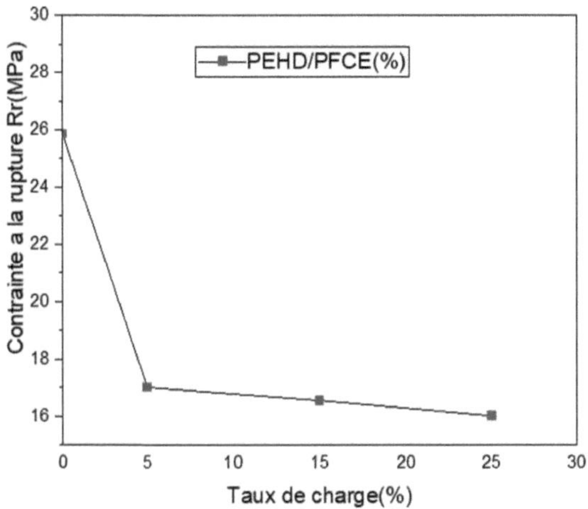

Figure V.3: Variation in stress at fracture of HDPE/PFCE composites as a function of

The PFCE load rate.The results show that the filler content has an influence on the mechanical properties of composite materials. The decrease in stress at break as a function of the increase in the percentage of PFCE filler indicates an increase in the stiffness of our material. A lower stress at break means a lower mechanical strength of the composite material. This is in fact due to the poor dispersion of the plant filler particles within the HDPE matrix.

c)   **Strain at break (elongation at break)**

This characteristic defines the ability of a material to elongate before failure when loaded in tension. The results of the strain to failure of HDPE/PFCE composites as a function of load content are shown in **Figure V.4**.

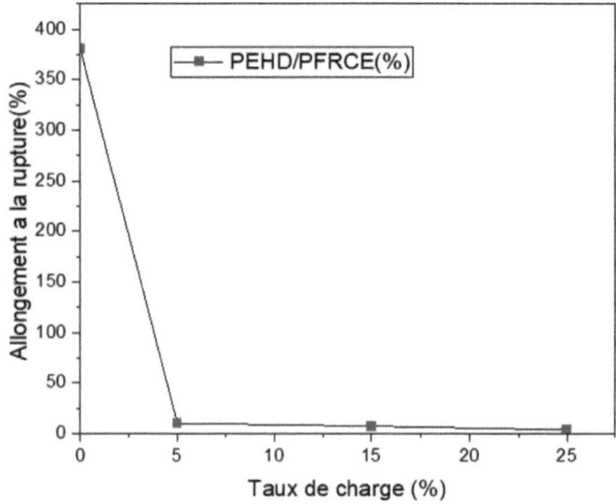

Figure V.4: Variation in elongation at break of HDPE/PFCE composites in PFCE load rate function.

The data shows that the elongation at break of the samples is reduced when the organic fibre content is increased, which means that the material has become stiffer by the addition of this filler, loses its flexibility, and its impact resistance increases due to the effect of the filler particles that make the material rigid.

d)   Young's modulus (modulus of elasticity)

Young's modulus is the initial slope of the stress-strain curve. The results of the modulus of elasticity of HDPE/PFCE composites as a function of the rate of the are shown in **Figure V.5.**

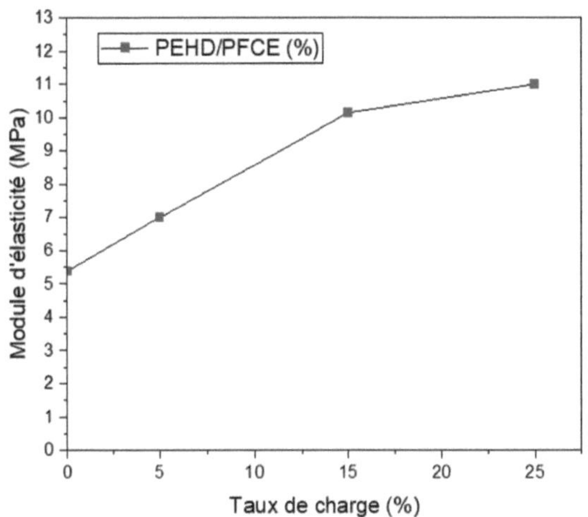

Figure V.5: Variation in modulus of elasticity (MPa) as a function of load rate PFCE.

Figure V.5 shows that the elastic properties are affected by the added organic matter content. The increase in elastic modulus is proportional to the increase in PFCE filler content, so the presence of the powder in the HDPE/PFCE composite increases the Young's modulus compared with pure HDPE. This composite becomes stiffer and loses its elasticity, and its impact resistance also increases, due to the action of the filler particles, which make the material solid.

V.4 Morphology by optical microscopy

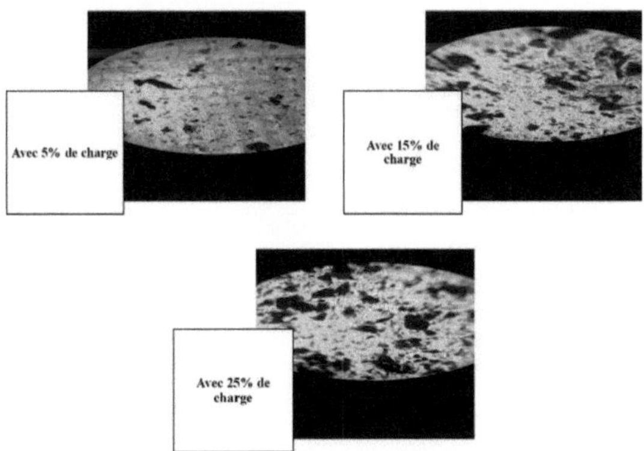

Figure V.6: The morphology of HDPE/PFCE composites at different loading rates.

From **Figure V.6** we can see that the agglomerates formed by the filler particles increase with the increase in filler rate due to the poor dispersion of the plant filler particles within the HDPE matrix. The phenomenon of agglomeration is due to the nature of the hydrophilic charge in a hydrophobic matrix.

V.5 Conclusion

From all the results, it can be concluded that the addition of prickly cactus leaf powder has little effect on the properties of HDPE. This is due to the poor dispersion of PFCE filler particles in the HDPE matrix and the total absence of filler/matrix interactions.

GENERAL CONCLUSION

The main objective of this study was to develop and prepare new composite materials based on a plant fibre (spiny cactus leaf), varying the percentage of this organic filler (0%, 5%, 15%, and 25%) in order to study their effects on the properties of HDPE, such as physical properties (density, water absorption), mechanical properties (tensile test) and morphological characteristics. The results of various characterisation methods have enabled us to draw the following conclusions: analysis of the density shows that its value decreases with increasing filler content (0.9541, 0.9418, 0.9326, and <0.93 for 0, 5, 15, and 25% respectively), which can be explained by the occupation of the free volume by less dense particles.

- The rate of water absorption is proportional to the increase in the percentage of filler in the composite, with HDPE/PFCE at 25% showing a high rate of absorption.
- The decrease in maximum stress as a function of the increase in load rate.

- Decrease in stress at fracture as a function of increasing load rate organic.

- The decrease in elongation at break as a function of the increase in the rate of the plant fibre.

- The increase in modulus of elasticity is proportional to the increase in the rate of PFCE charge.

- The agglomerates formed by the filler particles increase with increasing load rate.

OUTLOOK

In order to continue and develop this work, we have formulated the following perspectives:
- Chemical treatment of the fillers used (modification of the fibre surface).
- The study of rheological and thermal properties.

- The study of thermal properties by Differential Scanning Calorimetry (DSC) and thermogravimetric analysis (TGA).
- The use of an accounting agent, such as polyethylene maleic anhydride (PE- g-MA).
- Study the ageing of composites based on lignocellulosic fibres with different loading rates.

yes I want morebooks!

Buy your books fast and straightforward online - at one of world's fastest growing online book stores! Environmentally sound due to Print-on-Demand technologies.

Buy your books online at
www.morebooks.shop

Kaufen Sie Ihre Bücher schnell und unkompliziert online – auf einer der am schnellsten wachsenden Buchhandelsplattformen weltweit! Dank Print-On-Demand umwelt- und ressourcenschonend produzi ert.

Bücher schneller online kaufen
www.morebooks.shop

info@omniscriptum.com
www.omniscriptum.com